Wild Life LOL! ™
Llamas

You'll llove llearning about me!

SCHOLASTIC

Library of Congress Cataloging-in-Publication Data
Names: Grunbaum, Mara, author. | Children's Press (New York, N.Y.), publisher.
Title: Llamas/by Mara Grunbaum.
Description: [First edition] | New York, NY: Children's Press, an imprint of Scholastic Inc., [2020] | Series: Wild life LOL! | "Produced by Spooky Cheetah Press." | Includes index. | Audience: Grades 2-3 (provided by Children's Press)
Identifiers: LCCN 2019027496 | ISBN 9780531129791 (library binding) | ISBN 9780531132661 (paperback)
Subjects: LCSH: Llamas—Juvenile literature. | Lama (Genus)—Juvenile literature.
Classification: LCC QL737.U54 G78 2020 | DDC 599.63/67—dc23 LC record available at https://lccn.loc.gov/2019027496

Produced by Spooky Cheetah Press

Book design by Kimberly Shake. Original series design by Anna Tunick Tabachnik.

Contributing Editor and Jokester: Pamela Chanko

Printed in Heshan, China 62

SCHOLASTIC, CHILDREN'S PRESS, WILD LIFE LOL!™, and associated logos are trademarks and/or registered trademarks of Scholastic Inc.

1 2 3 4 5 6 7 8 9 10 R 29 28 27 26 25 24 23 22 21 20

Scholastic Inc., 557 Broadway, New York, NY 10012.

Photographs ©: cover, spine: Jose A. Bernat Bacete/Getty Images; cover speech bubbles and throughout: Astarina/Shutterstock; cover speech bubbles and throughout: pijama61/Getty Images; back cover: Milton Rodriguez/Shutterstock; 1: Tim Graham/Getty Images; 2: annamoskvina/iStockphoto; 3 top: Westend61/Superstock; 3 bottom: David South/Alamy Images; 4: Agatha Kadar/Shutterstock; 5 left: All-Silhouettes.com; 5 right: teka12/Shutterstock; 6-7: F1online digitale Bildagentur GmbH/Alamy Images; 8-9: Prisma by Dukas Presseagentur GmbH/Alamy Images; 10-11: Gannet77/iStockphoto; 12-13: Michel Gunther/Minden Pictures; 14 top: imageBROKER/Superstock; 14 bottom: Tim Graham/Getty Images; 15 top: susanna cesareo/iStockphoto; 15 bottom: Simon Balson/Alamy Images; 16-17: ANIMALS by VISION/Alamy Images; 18: Animals Animals/Superstock; 19 left: Janet Horton/Alamy Images; 19 right: Robert Llewellyn/Getty Images; 20-21: wanderluster/Getty Images; 22: OSTILL/iStockphoto; 23 top left: David South/Alamy Images; 23 top right: piccaya/iStockphoto; 23 bottom left: Flathead Beacon/Greg Lindstrom/AP Images; 23 bottom right: Luciana Zapatero Denegri/EyeEm/Getty Images; 24-25: Catmando/Shutterstock; 25 bottom right: Westend61/Superstock; 26 left: National Geographic Image Collection/The Granger Collection; 26 right: The Granger Collection; 27 left: TopFoto/The Image Works; 27 right: Jon12/Stockimo/Alamy Images; 28 left: Juniors Bildarchiv GmbH/Alamy Images; 28-29 top: Bob Wickham/Getty Images; 28-29 bottom: Iakov Filimonov/Shutterstock; 29 center: Anan Kaewkhammul/Shutterstock; 29 bottom right: David South/Alamy Images; 30 map: Jim McMahon/Mapman ®; 30 inset: Westend61/Superstock; 31: Ian Dyball/Shutterstock; 32: Luciana Zapatero Denegri/EyeEm/Getty Images.

TABLE OF CONTENTS

Ready for a llama-rama?

MEET THE SASSY LLAMA

Are you ready to be amazed and amused? Keep reading! This book will give you a lot to chew on.

FLEECED to meet you!

LOL!
What sound does a llama's doorbell make? Llama-llama-ding-dong!

At a Glance

Where do they live? → Llamas are originally from South America. Today they live on farms all over the world.

What do they do? → Llamas usually live in herds and spend much of their time eating together.

What do they eat? → Llamas can eat many different plants but mostly feed on hay.

What do they look like? → Llamas are hairy and have long necks, stocky bodies, and stubby tails.

How big are they? →

HINT: You're smaller. Check this out!

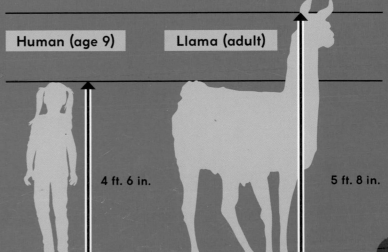

Human (age 9)   Llama (adult)

4 ft. 6 in.   5 ft. 8 in.

LLAMAS' ORIGINS

Llamas originally come from the Andes Mountains in South America. They can live in really high places!

THAT'S EXTREME!
Llama hair is usually 3 to 9 inches long, but can grow to 12 inches!

domesticated: tamed in order to live with people

Fast Friends
Llamas are friendly! They like being around other llamas, farm animals, and people.

LoL!
Why don't llamas laugh at jokes? They've HERD them all before!

Living Together
Llamas in the wild lived in groups called herds. Today, llamas live on farms all over the world. People still often keep them in herds.

Early Farmers
Thousands of years ago, the ancient Inca people **domesticated** llamas. They made clothing from llama wool and burned llama poop as fuel. They also ate llama meat.

A LLAMA'S BODY

Llamas are built to roam in high **meadows** and to climb mountains. They have good balance!

FAST FACT:
A llama's coat can be brown, black, gray, red, white, or multicolored.

THAT'S EXTREME!
An adult male llama can weigh more than 400 pounds. That's heavier than a refrigerator!

TOE-tally Cool

Llamas have two toes on each foot. This helps them balance on rocky ground. Each toe has a hard nail on top and a soft pad on the bottom.

meadows: grassy fields

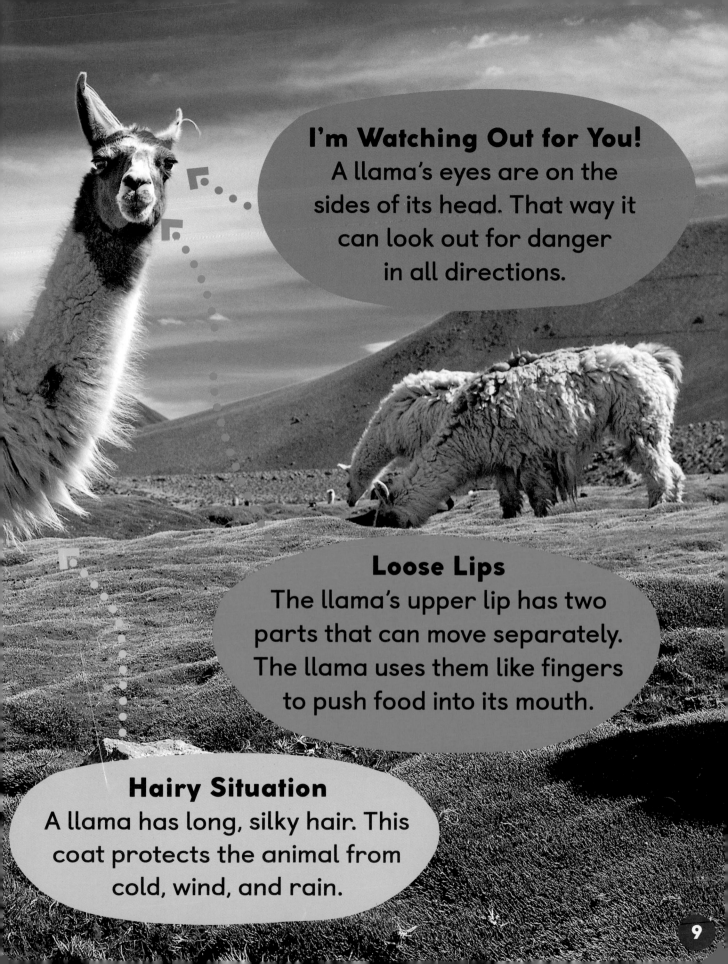

I'm Watching Out for You!
A llama's eyes are on the sides of its head. That way it can look out for danger in all directions.

Loose Lips
The llama's upper lip has two parts that can move separately. The llama uses them like fingers to push food into its mouth.

Hairy Situation
A llama has long, silky hair. This coat protects the animal from cold, wind, and rain.

WAIT! THIS IS NOT A LLAMA

This is an alpaca! Llamas and alpacas are very similar. People often get them confused! Here's how you can tell them apart.

Mini Me
Alpacas are about a foot shorter than llamas.

LOL!
What did the llama say when it was going on vacation? **ALPACA bag!**

All Ears
Alpacas have short, pointy ears. Llamas have long ears that curve like bananas.

Anybody have a comb?

Baby Face
Alpacas have short, fluffy faces. A llama's face sticks out more from its head.

Time for a Trim?
Alpacas have shaggier hair than llamas. Alpacas are often raised for their wool.

WACKY FACT:
An alpaca can produce up to 10 pounds of wool a year. That's enough for several VERY cozy sweaters.

LUNCHTIME

Llamas are **herbivores**. They spend much of their day **grazing** on plants. Here's how their bodies process all that grub.

LOL!
What did the llama say about the new girl at school? **I've never met HER-BIVORE!**

WACKY FACT: Llamas in a herd all poop in one big "potty pile." That keeps the rest of their grazing area clean.

THAT'S EXTREME! A typical farm llama eats about 5 pounds of hay per day.

1

Mixed Greens

Llamas have three-chambered stomachs. This system lets them eat grasses, leaves, and shrubs that are too tough for other animals.

herbivores: animals that eat only plants

> This is the last straw!

2 Food Processor

As a llama grazes, food collects in the biggest stomach chamber and starts to break down.

3 Chew on This

Hours later, the llama coughs up a glob of partly digested food called cud. It chews the cud to break it down further, then swallows it again.

grazing: feeding on grass that is growing in a field

LEARN TO SPEAK LLAMA

Llamas **communicate** by using sounds and body language.

I'm All Ears

When a llama is relaxed, its ears point forward. It pins its ears back to show other llamas that it's nervous or scared.

What Did You Say?

A curious llama makes a humming noise. If it senses danger, it warns the herd with a loud honking call that sounds like a squeaky toy!

communicate: to share information

Mind Your Manners

Llamas are famous for spitting when angry. They typically do this to other animals that are bothering them. They don't usually spit at people who treat them properly.

I'm So Excited!

When a llama is excited, it might jump up and down like a bunny. This happy hop is called "pronking."

GROWING THE HERD

Llamas start to **mate** when they are one to three years old. They typically do this in late summer. Here's what happens.

LOL!
What's a llama's favorite drink? LLAMA-nade!

1
Ready to Fight

Male llamas compete over females. They bite and try to knock each other over. The winner gets to mate with the female.

mate: to join together to have babies

FAST FACT: The word *cría* means "baby" in Spanish.

Back off, pal!

2
Baby on Board

Llamas are **mammals** that grow their young inside their bodies. A female llama is pregnant for about 11 months.

3
New Arrival

The female gives birth to her baby, called a cría, the next summer. Twins are very rare.

mammals: animals that produce milk to feed their young

HELLO, WORLD!

A female llama typically has one cría each year.

FAST FACT:
Llamas can live up to 25 years.

1

LOL!
What do you say to a spoiled baby llama?
Don't be a CRÍA baby!

Big Babies

Newborns usually weigh 15 to 30 pounds. Like all mammals, crías drink their mothers' milk. They start **nursing** right after they are born.

nursing: drinking milk from the mother's body

Born to Run

Crías are able to start walking within an hour of their birth. The young llamas spend a lot of time running and playing with other youngsters.

This Could be Cud

Most crías stop nursing by 10 months old. Many stop months sooner. Even as the cría begins to eat grass, it stays close to its mom. She teaches it how to find food.

LLAMAS AT WORK

Llamas are often used as pack animals. They carry food and supplies for people climbing up and down mountain trails.

WACKY FACT: Hospitals sometimes bring in llamas to help cheer up sick patients.

I CUD Sit Here All Day

If a llama is given too much to carry, it will simply sit down and refuse to budge. Once its load is lightened, the llama will start walking again.

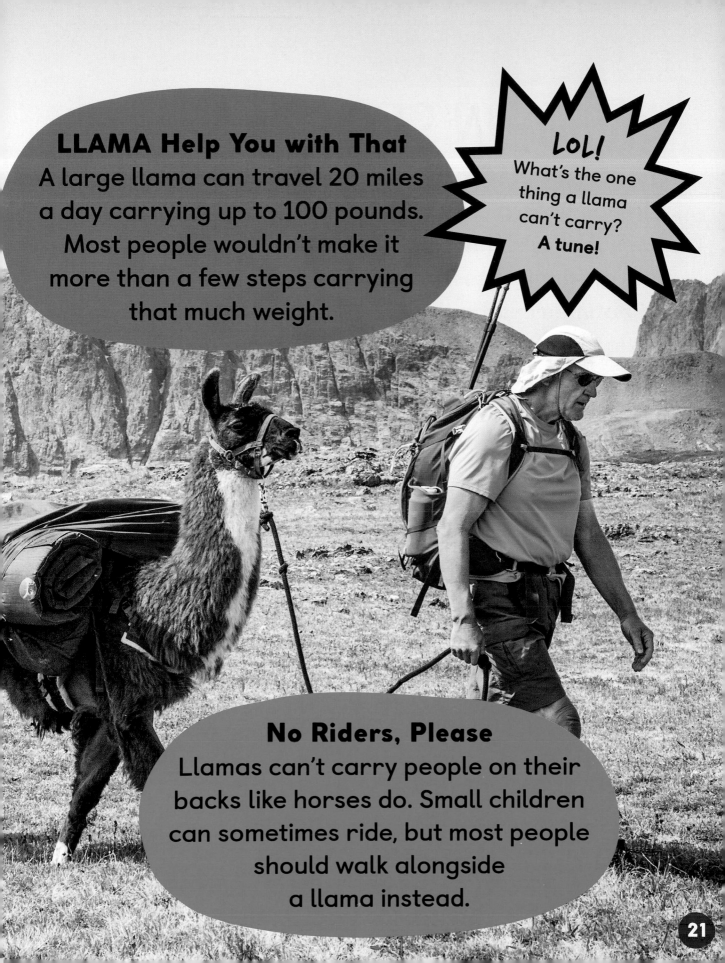

LLAMA Help You with That
A large llama can travel 20 miles a day carrying up to 100 pounds. Most people wouldn't make it more than a few steps carrying that much weight.

LOL!
What's the one thing a llama can't carry? **A tune!**

No Riders, Please
Llamas can't carry people on their backs like horses do. Small children can sometimes ride, but most people should walk alongside a llama instead.

LIVING IN STYLE

Llamas have plenty of personality! They make lots of funny faces. People sometimes celebrate them by dressing them in costumes and shaving their coats into unique styles.

WACKY FACT:
Unlike horses, llamas often lift both feet on the same side when they run. It's just another element of their style!

Red is definitely my color!

22

ANCIENT LLAMAS

Meet *Camelops*, the llama's ancient relative. This is how experts think it looked.

What's Your Name?
Camelops means "camel-face." This animal is related to both camels and llamas.

THAT'S EXTREME!
Camelops was about 7 feet tall at the shoulder. That's taller than most adult humans.

fossils: plants or animals from millions of years ago preserved as rock

Hmm . . . I guess I see the resemblance.

Home on the Range

Camelops and other llama relatives once roamed North America. Scientists have found **fossils** in states from Utah to Florida.

We'll Miss You

Camelops lived between 3.6 million and 11,700 years ago. These animals likely died out because humans hunted them and the **climate** changed.

LLAMA take a selfie and see.

climate: the typical weather of an area over a long time

LLAMAS AND PEOPLE

We have a long history together!

4000 B.C.

1500s

Llamas were domesticated by the ancient Inca people. They raised the animals for their meat, wool, and hides. Llamas were also important to the Incas' religion.

Spanish settlers arrived in South America. They used llamas to carry crops and other goods over the mountains. Llamas were sometimes called the "ships of the Andes."

1800s

The first llamas were brought to the United States by zookeepers. A wealthy businessman in California also bought a large herd as pets. Llamas grew in popularity over time.

Today

Llamas are bred on farms around the world. They are mostly raised as pets and pack animals, or sometimes for wool. Hikers and hunters can rent llamas to help carry their gear.

Llama Cousins

These are some of the llama's closest relatives. Most of them are from South America.

Our soft wool makes us popular farm animals.

vicuñas

We live in the wild.

alpacas

Please note: Animals are not shown to scale.

The Wild Life

Llamas no longer live in the wild. But guanacos, their closest relatives, do. Check out the red areas on this map of the world. They show that guanacos still live where llamas came from: the mountains of South America. Up to 50 million guanacos once roamed this area.

South America

GUANA-CO meet my family?

habitats: the places where plants or animals make their homes

Helping the Llama Family

In the 1800s, people hunted many guanacos. People have also turned some guanaco **habitats** into sheep farms. Today, only about 2 million guanacos remain in the wild.

In the 1970s, authorities began protecting guanacos from hunting. Some countries promised to preserve wild areas where guanacos and other animals live. Over time, guanaco numbers have started to rise again.

What Can You Do?

Talk to an adult about supporting nonprofit organizations such as the Wildlife Conservation Society. This organization works with farmers to save guanaco habitats. Tell your friends what you've learned about llamas and their relatives. The more people care about animals, the more they'll want to protect them.

INDEX

ABOUT THE AUTHOR

Mara Grunbaum is a science writer who loves learning about odd and fascinating animals. She lives in Seattle, Washington, with Zadie, the world's smartest cat.

Llater!